Jorge Alfredo González Pérez
Augusto Manuel Kahali Petuleinge

Design of actions to improve the agricultural marketing process

Jorge Alfredo González Pérez
Augusto Manuel Kahali Petuleinge

Design of actions to improve the agricultural marketing process

Beans on the informal market

ScienciaScripts

Imprint
Any brand names and product names mentioned in this book are subject to trademark, brand or patent protection and are trademarks or registered trademarks of their respective holders. The use of brand names, product names, common names, trade names, product descriptions etc. even without a particular marking in this work is in no way to be construed to mean that such names may be regarded as unrestricted in respect of trademark and brand protection legislation and could thus be used by anyone.

Cover image: www.ingimage.com

This book is a translation from the original published under ISBN 978-620-0-01014-8.

Publisher:
Sciencia Scripts
is a trademark of
Dodo Books Indian Ocean Ltd. and OmniScriptum S.R.L publishing group

120 High Road, East Finchley, London, N2 9ED, United Kingdom
Str. Armeneasca 28/1, office 1, Chisinau MD-2012, Republic of Moldova, Europe
Managing Directors: Ieva Konstantinova, Victoria Ursu
info@omniscriptum.com

Printed at: see last page
ISBN: 978-620-8-51114-2

DEDICATION

- To my family, in particular my parents, for everything they have done for my life in every way.
- I hold my brothers in high esteem and love.
- I would like to dedicate this to my friends, sports activity groups called "Los Amigos", for their invaluable support during this odyssey.
- To my fellow students of Agricultural Engineering at the Polytechnic Institute of Ondjiva.
- To the entire scientific community, as this work will serve as a source of knowledge, especially for those in the field of agricultural engineering.

THANK YOU

- I thank GOD, the almighty creator of heaven and earth, for the health, patience and wisdom he gave me to do this work, and, above all, for the life he gave us until then.
- MANDUME YANDEMUFAYO UNIVERSITY, IPO, for granting me the privilege and honour of a truly fruitful training.
- To my excellent teachers, administrative staff, cleaning assistants, caretakers, among other employees attached to this Higher Education Institution, who directly and indirectly contributed to my training.
- I am very grateful to my classmates and classmates for the excellent and unforgettable moments we had together.
- I am very grateful to my eternal colleague "Manuel Francisco", may God have him in all his glory and may the moments spent in his company never be forgotten, as they will remain forever in my memory.

SUMMARY

In order to propose a set of actions to improve the bean marketing process, this work was carried out in informal markets in Oshomokyo and Germany between February and May 2022. This study was considered exploratory-descriptive in nature. characteristics, it was possible to describe a population, to know the behaviour and qualities of consumers in relation to the products and to have knowledge about the trend of supply and demand. A non-experimental design was used, using observation and scientific research. Field data was collected in informal markets where 8 randomly selected consumers (customers) and 11 vendors were interviewed, 6 in Germany and 5 in Oshomokyo. The information collected was processed using tables and graphs in Microsoft Word and Excel for Windows 10. According to the results: Bean marketing either operates under the producer-cooperator or middleman-wholesaler-consumer sales model. In German and Oshomokuyo markets, 92% of consumers surveyed stated that bean food offerings are good, but are not sold at different prices, and Oshomokuyo is where the largest variety of beans is sold. The situations found in FOFA in both locations presuppose similar reactions, but it was agreed that a weakness or threat in Germany here will not necessarily be in favour of Oshomokuyo. The actions outlined above should be carried out in different ways, but the most efficient way is through government action projects with a clear intentionality in their implementation.

Keywords: bean, action plan, informal markets

INDEX

INTRODUCTION

I. INTRODUCTION

The bean (*Phaseolus vulgaris* L.) is a plant that has been cultivated by humans for thousands of years. Its origin is in the Americas, with the oldest archaeological evidence dating back around 8-10 thousand years in South America (Kaplan *et al.,* 2016). Regarding its domestication, it is argued that there are divergences.

Worldwide, beans (all species taken together) are grown in more than 100 countries. Brazil and India are the largest producers and together account for more than 35% of world production. Around 23 million tonnes are produced annually, with an occupied area of approximately 30 million hectares (FAO, 2020).

The volume of beans traded on the international market is considered low (FAO, 2019). Considering that only 14% of world production is destined for export. One of the reasons given for the low international trade is the large variety of bean types and the differences in eating habits among consuming countries.

The "marketing of agricultural products" is the whole process from the time the product leaves the producer's or entrepreneur's company, farm or holding to the hands of the consumer, Fisher and Navarro (2018). The same authors state that marketing means not only the action of buying or selling, i.e. the change of ownership of the good, but also the physical aspects of transport (change of place), storage (change in time), packaging and processing (form of exchange).

Marketing is the main problem faced by the agricultural sector, as there are no companies dedicated to this activity, and on the other hand, the rate of intermediaries has been increasing, generating inequality between direct prices to the producer and to the consumer. Finally, this makes it necessary for small producers to find new alternatives to market their products, in order to form leading companies dedicated to the marketing of agricultural products at fair prices and achieve the integration of different producers to improve the quality, quantity and prices of their products. products in the market, (NEVES E CASTRO, 2017).

In Angola, beans are grown by small and medium-sized producers, in diversified production systems (including fully mechanised ones, from sowing to harvesting) and in all regions of the country, ranging from production for the farmers' own subsistence, to the businesses of large producers. level, with a considerable technological burden invested. This difference is mainly because culture can be produced in any region of the country, regardless of the technological level, and also because consumption is widespread throughout the country, both in rural and urban areas. (Diario de Angola, 2021).

According to Moreira (2021), Angola has an annual bean deficit and in 2019/2020 Angola stood at just 634,205 tonnes, not a very significant percentage of world production of 26.84 million tonnes per year. In the same period, 344,762 tonnes of beans were destined for human consumption in 2019/2020, at the same time as 46,580 for seeds and 59,691 for other uses, registering losses estimated at 13,111 tonnes.

Bean cultivation is dominated by the family farming sector, which employs 2,936,198 families, covering a total area of 5,325,952 hectares per year, while the business sector brings together a total of 8,826 small, medium and large farms and a total area of 493,990 hectares per year (Jornal de Angola, 2021).

According to the same source, "the cultivation of beans is one of the most profitable, because it is a product that is part of the diet in almost the entire national territory. Its production is concentrated in the provinces of Bié, Huambo, Malanje, Uíge, Cuanza-Sul, Huila and Benguela, but all the provinces have potential. It is produced during three seasons of the year, but it is also possible to invest in greenhouse cultivation, which allows harvesting at any time of the year.

In the province of Cunene, specifically in the commune of Ondjiva, beans are an important part of the local diet, so the levels of commercialisation vary and present difficulties inherent to the process. Unfortunately, at the moment there is not enough information to know the behaviour of this product in informal markets.

Based on these elements described above, the following scientific problem was formulated: How does the agricultural commercialisation process of beans (*Phaseolus vulgaris* L.) develop informal markets in the municipality of Ondjiva in the municipality of Cuanhama?

Based on the scientific research problem, the following **hypothesis** is basically defended**:** The efficiency of marketing could improve the agricultural bean (*Phaseolus vulgaris* L.) if the factors that affect this process in informal markets in the Commune of Ondjiva in the Municipality of Cuanhama were known.

In order to respond to the problem posed, the following **General Objective** was defined:

to design actions to improve the process of agricultural marketing of beans (*Phaseolus vulgaris* L.) in informal markets in the municipality of Ondjiva in the municipality of Cuanhama.

In order to achieve the general objective, the following **specific objectives** were formulated**:**

- ✓ To characterise the development of the agricultural commercialisation process of beans (*Phaseolus vulgaris* L.) in informal markets in the municipality of Ondjiva in the municipality of Cuanhama.
- ✓ To identify the factors that affect the agricultural marketing process of beans (*Phaseolus vulgaris* L.) in informal markets in the municipality of Ondjiva in the municipality of Cuanhama.
- ✓ Propose a set of actions to improve the sale of beans (*Phaseolus vulgaris* L.) in informal markets in the commune of Ondjiva in the municipality of Cuanhama.

II. LITERATURE REVIEW

2.1 Taxonomy of the common bean (*Phaseolus vulgaris* L.)

Common fenjo belongs to the genus Phaseolus and its taxonomic location is: Rainho: Plantae; division: Magnoliophyta; class: Magnoliopsida; subclass: Rosidae; order: Fabales; family: Fabaceae; genus: Phaseolus and species: *Phaseolus vulgaris* L.

2.2 General aspects of common bean (*Phaseolus vulgaris* L.)

The origin of *Phaseolus vulgaris* was a much debated topic among historians, many believe that this crop originated in Europe and others point out that it originated in some regions of America, from where it was distributed to other continents, and after further studies. Detailing its origin, there were sporadic appearances in American countries, Peru, Mexico, Guatemala, Chile and Bolivia, which leads to the conclusion that it is of American origin. This certainty is based on data obtained from 1500 isolated points that appear in different descriptions and references (González, 2018).

Among the food legumes, beans are one of the most important species for human consumption. Their production covers diverse agro-ecological areas. This legume is cultivated practically all over the world (FIEP, 2017).

2.3 Nutritional and economic importance of the crop

The common bean known as Frijol, Frijol riónes, Frijol Poroto and Caraota is a crop of great importance in human nutrition due to its high nutrient content. In Latin America it is an essential component of the diet as it is an important source of protein (Socorro and Martín, 2009).

According to Castiñeiras (2011), beans are the legume that has been the subject of most studies in Latin America, as it is the main source of protein, in addition to being an important part of the population's eating habits. Scientific advances corroborate the need to incorporate and maintain this food in the conventional diet, due to its proven nutritional and medicinal values. It also stands out for its protein content, which is why it is included in food guides in the group of foods characterised by their protein content.

Beans provide vitamins such as B-complex and folic acid. Vegetables are rich in dietary fibre and minerals such as calcium, iron, copper, zinc, phosphorus, potassium and magnesium. Beans are one of the main sources of soluble fibre in the common diet and help reduce cholesterol (Silva, 2018).

2.4 Morphology of common bean (*Phaseolus vulgaris* L.)

The bean is a herbaceous plant, with a relatively short biological cycle, of annual character, of variable size and habits, with varieties of determinate and indeterminate growth (small shrub, climber) as described (Socorro a Martín, 2009).

In its morphological constitution, the bean consists of the following organs:

Root: The root system consists of a main root as well as a large number of secondary roots. When germinating, it grows rapidly, its active layer developing between 0.20 - 0.40 cm deep and within a radius of 0.15 - 0.30 cm. With numerous lateral branches, this crop has the ability to fix atmospheric nitrogen through symbiosis with bacteria of the genus Rhizobium by forming nodules on its roots, Gepts and Debouck (2015).

Stem: The stem is made up of nodes and internodes which vary in size, and from each node a leaf emerges; its height depends on the growth habit (determinate or indeterminate). They are called determinate when they reach a short height (0.20-0.60 cm) and have an inflorescence at their tip, while the indeterminable ones can be two to ten metres long and do not have an inflorescence at their terminal bud (Ferreira *et al.,* 2012).

Leaves: the leaves, in turn, are alternate, composed of three leaflets (two lateral and one terminal or central). The leaflets are large, oval, with pointed and pointed ends (Unifeijão, 2011).

According to Wander (2007), "The leaflets are oval or rhomboidal in shape. There is a midrib and a system of branched veins throughout the blade area. The leaves are alternate, trifoliate and dark or light green in colour. The shape of the leaflets is varied: oval, deltoid and wedge-shaped. There are also trifoliate leaves.

Inflorescence: they can be terminal (only present in certain growth varieties) and axillary, which are present in both growth habits. The flowers have five unequal petals: one standard, two fused ones forming the keel and two "wings". The flower is symmetrical and can be of different colours: white, pink, yellow, violet (Diniz, 2018).

Fruit: it is a vegetable commonly known as pod, with an enlarged shape, which may have different colours such as: cream, brown, purple. The pod contains three to nine seeds, but the normal number is five to seven, of reniform shape, although they can also be round, ovoid, elliptical, small, almost square or ovoid (Muñoz and Singh, 2003).

Seed: depending on the colour, you can find uniformly coloured grains, for example, black, red and white, you can also find them in two colours with different variants within this group, and finally even in three different colours, the state of physiology Maturity, or term of growth of the grain, is reached when a moisture content of 52 to 54% is obtained on average. The colour of the grains is green at the beginning of their growth, until they reach a moisture content slightly above or very close to 60%; thereafter, the grains acquire the characteristic colours of each cultivar, until they obtain their definitive colour at physiological maturity (Diniz, 2018).

The seeds of this crop have the property of rapidly losing moisture once ripe, being able to be stored without major difficulties, since their teguments are very impermeable, although their thickness is a characteristic that depends on the variety and type of bean (Socorro and Martín 2009)).

2.5 Marketing alternatives or strategies

"Alternative" is defined as a procedure, mechanism, method or option by which a producer can sell or influence the conditions of sale of his product. The main alternatives available to a producer are; cash sale at harvest time, pre-harvest sales contract, speculation bets, authorised price, fixed price, average price or joint sale (Mendes *et al.,* 2007).

2.5.1 Principles of the strategy

Agricultural production as an economic activity and associated with a multitude of variables that condition both technological results and profitability, also requires the adoption of a strategy, both those of a technical nature (physical and biological aspects) and those of an institutional and human nature, considered exogenous or endogenous to agricultural property (Neves, 2010). Furthermore, according to this, the author points out that the agricultural marketing strategy is the set of principles, objectives, actions and priorities for the development of agricultural trade, supported by private initiative, market forces and the regulatory and facilitating role of the State. it must comply with the following principles:

- Fulfilment of fundamental choices;
- Definition of trade as an activity essentially based on private initiative and the necessary link between production and consumption;
- The need to promote and modernise the commercial network and related services;
- The need to promote the marketing process taking into account economic interests, particularly in the protection of agricultural production and trade.

2.6 Proposed actions. Definition

A pattern or plan that integrates an organisation's main objectives, policies and sequence of tasks into a coherent whole. The authors relate a set of well-formulated actions to a unique and viable position obtained through its own abilities, but controlling its internal deficiencies and the measures taken by intelligent opponents (Senaes, 2013).

2.7 General information on the marketing process

Agricultural marketing plays an extremely important role in the national economy, constituting the main, if not the only, source of income for people in rural areas, where the majority base their livelihoods on subsistence farming. Marketing is one of the drivers of producer/farmer-market linkages in terms of rural-urban economic relations (Agribusiness, 2014).

For Barros and Martines (2017), agricultural marketing refers to a set of functions or activities of transformation and utility addition through which goods and services are transferred from producers to consumers. Thus, in the broadest definitions, marketing of goods, which is therefore also a production process. In addition, marketing comprises activities that result in the transformation of goods, through the use of productive resources: capital and labour acting on agricultural raw materials.

Marketing is based on the transit of goods to a final destination, as long as standards and procedures are met and customer satisfaction is achieved, specifically in agricultural products that have a more demanding market, where the quality of the products must be highlighted. (pizza ygalés, 2018).

Since ancient times, man has had to use exchange as the first form of trade in order to survive and then develop. In the semantic order, commercialisation is synonymous with marketing in English or marketing in Spanish, the term comercial, which means to negotiate the buying and selling of something (Pinho *et al.,* 2014).

Agricultural marketing literature presents several definitions with varying degrees of breadth. In a narrower sense, marketing is defined as the set of operations that begin with the creation of agricultural products and culminate in their use at the final consumer level. In this context, marketing means the continuation of the production process (Portero, 2006).

Agricultural marketing is not simply selling the production in a continuous and organised process of routing agricultural production along a marketing channel or system, where the product undergoes transformations, differentiations and added value. The facilities (utilities) that agricultural products undergo are possession, form, time and place, adapting them, in this way, to the taste and preference of the final consumers (Neves,2010).

Agricultural marketing is presented as the most complex activity among those involving the agricultural system, as it is the moment when production assumes the status of a commodity. This condition reflects the dynamics in which the integration of markets, comprising different segments and sectors, almost appropriates production and begins to impose quantity and quality goals, forming chains, networks or productive arrangements (José, 2015).

2.8 Terms used in the marketing process

According to Navalar *et al.,* (2010) the following are used:

Marketing: Marketing comprises the set of activities in the transfer of goods and services from the initial point of production until they reach the final consumer.

Trade: the process of buying and selling or exchanging goods of different kinds.

Market: should be understood as the place where the forces of supply and demand operate, through sellers and buyers, so that the transfer of ownership of goods occurs through buying and selling transactions.

According to Guzmán (2015), several methods operate in the marketing activity that facilitate this activity, among which the following can be highlighted:

- Correct marketing;
- Incorrect marketing;
- Wholesale and retail marketing

2.9 Types of marketing

According to Garófalo and Carvalho (2014), domestic consumption or micro-marketing consists of obtaining goods and services from the producer-consumer or market-consumer; and that this distribution can be carried out directly or indirectly so that customers can purchase them, i.e. they can decide to sell products or services to the end user.

Micro-marketing is the performance of those activities that aim to achieve the objectives of an organisation by anticipating customer needs and directing a flow of goods and services that satisfy the needs of the producer to the customer.

The same author considers that it can therefore be said that external consumption or macro-marketing is an export, i.e. it is the shipment of goods and services to another part of the world for commercial purposes. The shipment can be done through different means of transport, whether by land, sea or air.

2.10 Marketing channel

Marketing channel is the path that goods follow from the producer to the final consumer. It is the sequence of markets through which the product passes, under the action of various intermediaries, until it reaches the region of consumption. The marketing channel shows how intermediaries are organised and grouped to move production to consumption. The nature of the various intermediaries and agencies that perform marketing services for a product, as well as the arrangement and organisation of the market mechanism (franco, 2011).

The same author considers that, in this method, the human element receives special emphasis. Intermediaries are individuals or commercial organisations that specialise in carrying out various marketing functions, related to buying and selling activities, as goods pass from producers to consumers. Intermediaries of direct interest in the marketing of food products can be classified as follows:

the) commercial intermediaries: wholesalers, retailers or retailers and speculators;

b) intermediary agents: brokers and commission agents;

w) auxiliary or instrumental organisations;

d) processing industry.

Intermediary traders have title to the goods and therefore own the products they handle. They trade for their own profit, securing their income from the margin between the buying and selling prices of the goods they sell. Wholesalers sell to retailers and other wholesalers and manufacturers, but do not sell significant quantities to the final consumer.

Retailers buy products from wholesalers for resale to the end consumer. They constitute the largest group among marketing agencies. Intermediary agents, as they are called, act only as representatives of their customers. They do not hold title and therefore do not own the goods they sell; their income is represented by fees and commissions on the volume of sales they make (Senaes, 2017).

For Figueiredo *et al.* (2004), commission agents or inspectors generally have great authority over the goods, being responsible for their movement and setting the conditions of sale and deduction of fees. Brokers do not regularly have physical control over the goods they handle, closely following their clients' orders. Their powers in negotiations are less than those of commission agents.

Speculative intermediaries are a group that appropriates products, with the aim of profiting from short-term price fluctuations. The buying and selling activity is usually carried out at the level of the market channel. In competition with other intermediaries, these agents contribute to maintaining an appropriate price structure (Sabourin, 2010).

According to Ledesma (2017), the classification of marketing channels is based on their length and complexity. The most common types are:

- The producer sells directly to the consumer, an example is what happens with traders who are producers who sell their production directly to the consumer.
- Transactions are carried out by intermediaries. In this case, the marketing channel can be of varying complexity, depending on the number of transactions and thus the number of people involved. As the economy develops and the specialisation of the activity intensifies, the channel tends to become more complex.

The relationship between consumer supply and demand is widely identified as the main determinant of market price. However, the above analysis of market levels and marketing agents showed that consumers and producers are separated by many intermediaries that allow the transfer of agricultural production to final consumers. This intermediation results in marketing costs that will be higher or lower depending on the levels of intermediation (Rivas, 2007).

2.11 Product. Concept

The product is a set of attributes that the consumer considers a certain good to have in order to satisfy his needs or desires. According to a manufacturer, the product is a set of physical and chemical elements combined in such a way as to offer the user possibilities of use. It is that, desired or undesired, which a person or company realises in an exchange; the basic objective of purchasing decisions is to receive the tangible or intangible benefits associated with a product; tangible and intangible aspects include service, retailer image, manufacturer reputation and social status associated with a product; a company's product offering is the crucial element in any marketing mix.(Neves *et al.,* 2013).

Consumers' interest in buying locally produced agricultural products, i.e. in their region of residence or country, has been growing. According to the perception of quality, the demand for fresh produce, the short distance between products and consumers are some of the essential factors for the demand for local products and consumers are more likely to pay a premium price for the products (Singer, 2016).

2.12 Promotion

According to Kaplan and Cooper (2018), they are defined as one of the fundamental marketing instruments with which the company intends to convey the qualities of its product to its customers, so that they are encouraged to purchase it; therefore, it consists of an information transmission mechanism. The promotion of a product is the set of activities that attempt to communicate the benefits provided by the product and persuade the market to buy from those who offer it.

2.13 Marketing of agricultural products

For Kotler (2000), Agricultural Marketing has gained importance due to:

- Increased consumer power;
- Industrialisation of agribusiness;
- Increased regulation of agricultural production;
- Progressive globalisation of agriculture;
- Marketing;
- A personal attitude and a company culture.

2.14 Factors influencing the marketing process in the country.

Coordination and integration of the segments of production, storage and conservation (concentration of supply, in quantity and quality), eventual industrial processing, marketing and distribution of agricultural products. This is recognised by all as one of the most important limitations for the full development of Angolan agriculture, since it is accepted that its correct development will immediately trigger an incentive for all forms of agriculture which, for lack of them, have often opted not to produce, so as not to have to lose production due to lack of flow (MINADER-FAO, 2003).

This marketing must be two-way to be truly effective, as it must not only be able to bring factors of production and other elements closer to the farmer, but also have the capacity to sell the products produced in the region where they exist. The prices at which factors are sold and products purchased must be adequately considered so as not to hinder the success and development of more distant regions, which will naturally also be linked to the ease of road or rail transport that the products in question may merit (Barros, 2007).

In this way, it can be seen that agriculture has problems "inside" and "outside" the property, but it is undoubtedly outside the reach of rural producers where most of the problems that affect the economic-financial result occur. with adverse social consequences (Rivas, 2007).

The identification of new products that can help diversification and create added value in the domestic or external market will also be very important to overcome this situation, as the sector is raising awareness and mobilising private entrepreneurs to get involved in domestic seed production, mainly of food crops such as maize, beans, rice and cassava (Mamaot, 2013).

The country needs to make an effort to address the assumptions for agricultural development. It is necessary to create alliances with companies with experience in the sector and to take advantage of the knowledge of the countries in the region (Zimbabwe, South Africa and Zambia) that produce the seeds they consume. The country invests foreign exchange to buy seeds that are not even enough to cover 50% of the country's needs in the sector. In addition to seed production, it is necessary to think about the issue of fertilisers, pesticides, tractor assembly and watering (CGA, 2010).

2.15 Market information

According to MINADER, (2020) the function of market information refers to the collection, interpretation and dissemination of data in order to facilitate "marketing". An important characteristic of information is that it must be current and reliable. There are three types of information:

(a) purely informative or newsworthy;

(b) market analysis (prospects) and,

(c) advertising.

The "informative" type only contains data on prices, supply conditions, stock volumes, weather, etc., without any analysis or commentary on the market situation.

The "analytical" type goes beyond news, because it presents explanations (reasons) for the current trend and makes predictions of this trend. In this case, in addition to data on the relevant variables, it is necessary to analyse these data using statistical and economic models . In this case, knowledge and factors linked to agricultural supply and demand are needed.

This same source goes on to say that among the relevant variables on the demand side are the following indicators: domestic population, disposable income level, employment level, per capita consumption, changes in tastes and preferences, prices of substitute goods, external demand and special government programmes. On the supply side we have: planting intentions, price expectations, competitive product prices, expected productivity, area available for planting and adoption of technology packages.

Another type of market information can be obtained through advertising, which takes two forms. The first is called "generic", and can be carried out by the government or by a group of companies with the aim of increasing consumption of a product, without a specific brand.

2.16 Market research

In a purely business context, research related to changes in consumer preferences is important in determining business policy. Thus, research on packaging in terms of shape, size, colouring, consumer behaviour, sales forecasts in a certain region, research aimed at reducing 'marketing' costs, better means of communication for advertising, among others, are useful information in the long term. success of the company (Junta Provincial de Povoamento, 2015).

In this regard, Diniz (2018) reports that, more than just rural economics, market research is important in the following areas:

a) Demand and expenditure studies;

b) Offer studies;

c) Marketing cost analysis;

d) Analysis of marketing margins;

e) Agricultural price analysis;

(f) Market structure studies

III. MATERIALS AND METHODS

3.1 Location of the work

In order to propose a set of actions to improve the bean marketing process, this work was carried out in informal markets in Oshomokuyo and Alemania between February and May 2022. The province is located between latitudes 15° 10' 00" North, 17° 24' 00" South and longitudes 13° 02' 00" West 17° 23' 00" East (Ministry of Territorial Administration, 2019).

3.2 Type and method of research

According to the nature and characteristics of this study, it was considered exploratory-descriptive because due to its characteristics it is possible to describe a population, understand consumer behaviour and product qualities, and gain knowledge about supply and demand trends.

As an empirical research method, a non-experimental cross-sectional design was used, with the use of observation and scientific research, to capture information at a single point in time, with the purpose of describing variables and analysing their incidence and interrelationship at a given time.

3.3 Methodology used

The methodology used was Silva (2010), a survey, defined as an in-depth research technique, where information is obtained through the collection of primary data, completing a qualitative research in terms of data collection and quantitative in relation to the interpretation of these data. Initially, we seek to theoretically ground the propositions presented in this paper through a review of the literature on the subject. The information gathered from the literature review provided support for the construction of a questionnaire used in the data collection.

The questionnaire developed contained open and closed questions, which served as a guide for conducting interviews with producers and rural traders in informal markets. Subsequently, a field survey was carried out in order to collect information, with the objective of finding out, among other aspects, the operational structure, the flow of information and the marketing systems used by producers in the locality and the surrounding municipalities that bring the product. goods.

Regarding data collection in the field, 8 randomly selected consumers (customers) and 11 sellers were interviewed, 6 in Germany and 5 in Oshomukuyo. As a strategy for selecting rural producers and bean traders to be interviewed, two marketing points with wide regional insertion (two informal markets) were defined to make the products available.

3.4 Design and sampling

The type of sampling used was simple random, calculating the sample size, with a population of 8 producers and 11 sellers, with a margin of error of 5% and a reliability of 95% in standard units corresponding to the valued sample (1.96). The formula mentioned by Piza and Welsh (1968) was used.

Z2 because north

north= $\frac{\text{Z2 because north}}{\text{0,052 (N- 1) + Z2 because}}$ Where:

n = sample size to be determined.

N = size of the universe (Total number of elements in the population).

p = probability of the event occurring (50%).

q = probability that the event does not occur (50%).

E= maximum plausible deception as a function of the desired level of confidence (this deception is squared).

E= 5% (0,05).

Z = Confidence coefficient for a given probability level. This coefficient squared = 95% corresponds to a value Z= 1.96.

north = $\frac{\text{(1,96) 20,5*0,5*16}}{\text{0,052 (16 - 1) + 1,96 2 (0,5 * 0,5)}}$

north= 33.65

n= 34 tracks approximately.

3.5 Methodological steps

In order to avoid errors in the research process, different methodological steps were planned, which were as follows (see figure 1).

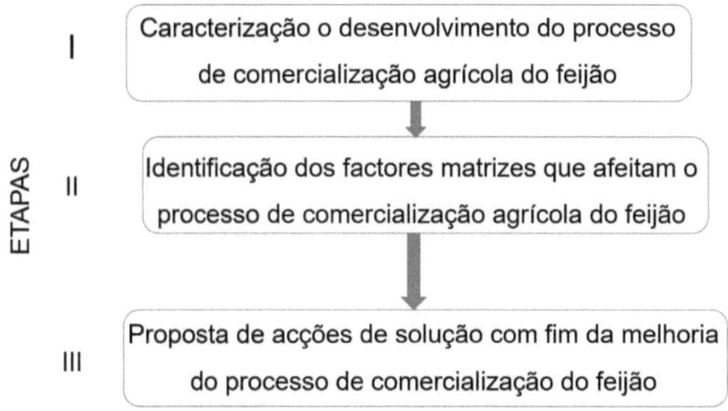

Figure 1. Diagram of the steps involved in dog research

3.6 Characterisation of the development of the process of agricultural commercialisation of the kidney bean (*Phaseolus vulgaris* L.) in informal markets in the Commune of Ondjiva in the Municipality of Cuanhama.

Due to the characteristics of the market structure, the methodology of Eterno and Elenor (2015), which shows how the bean marketing process is, for this part a diagram was created explaining the aspects that most influence the marketing process in the locality.

For this variable of establishments selling beans, different bean sellers were counted in both markets and the results are represented in the bar chart.

In the case of the change in bean supply-demand in local markets, a survey was conducted where the responses of buyers were collected and the responses were plotted in a table to show how the different variations occur according to levels of bean sales and presence.

3.7 Identification of the matrix factors that affect the marketing process. The factors affecting the agricultural marketing of beans (*Phaseolus vulgaris* L.) in informal markets in the Commune of Ondjiva in the Municipality of Cuanhama.

3.7.1 SWOT analysis variables

Paraces. Maintains a high level of performance, generating clear and present advantages or benefits, with attractive possibilities for the future.

Weaknesses or weaknesses: Means a deficiency or shortcoming, something in which the organisation has low levels of performance and is therefore vulnerable.

Opportunities. These are those circumstances in the environment that are potentially favourable for the organisation and may be changes or trends that are detected and that can be used advantageously to achieve or exceed the objectives.

Threats. These are environmental factors that result in adverse circumstances that jeopardise the achievement of established objectives. They can be changes or trends that appear suddenly or gradually.

SWOT analysis

This study was carried out according to the criteria of Rodríguez (2000) and Ramírez (2007), who mentioned that it consists of carrying out an evaluation of the strong and weak factors that together diagnose the internal situation, it is a tool that can be considered easy and allows an overview of the strategic situation of a given organisation to be obtained.

3.8 Proposal of solution actions to improve the sales process of beans (*Phaseolus vulgaris* L.) in informal markets in the Commune of Ondjiva in the Municipality of Cuanhama.

The training methodology described in this document is participatory. It suggests the development of group sessions, reflections, presentation dynamics, practical work and evaluation, with the aim of dynamising and strengthening the learning process and encouraging new attitudes and values such as: initiative, creativity and discipline in producers.

In addition, it will be structured in such a way that it can guide the facilitator in a simple way to achieve the objectives of the meeting. The topics covered in this document include a methodology sheet containing the following information:

- Objectives
- Expected target
- Method of monitoring activities.
- Persons responsible or involved

3.9 Organisation and statistical processing of information.

The information gathered was done through coding.research questions, creation of figures, tables, diagrams and data matrices.

The variables that formed the data matrices are indicated in the results and discussion of the paper. The results of the research were processed using tables and graphs. The information was analysed using the Excel package for Windows version 10.

IV. RESULTS AND DISCUSSION

4.1 The development of the agricultural commercialisation process of beans (*Phaseolus vulgaris* L.) in informal markets in the commune of Ondjiva in the Municipality of Cuanhama.

4.1.1 Market structure

According to the research carried out and shown as a representation of the process in figure 2, it is possible to observe the marketing channel identified for beans in the locality....

The sales chain operates under the sales model: producer-copier or intermediary-wholesaler and consumer. The intermediary buys the product in the production areas, on property or in the nearest regional markets, determining the price based on the supply and the price expectation to be obtained in the wholesale centres.

Figure 2. Interpretation of the bean supply chain

A market is the space or context where the exchange, buying and selling of bean services takes place between a buyer who demands them and a seller who offers them. Informal traders in the city belong to a competitive market, sometimes called a perfectly competitive market, where there are many buyers and sellers and the goods offered by the different sellers are basically the same. Competitive buyers and sellers must accept the price determined by the market and are therefore said to decide the price.

According to the analysis, a direct marketing channel is identified, where producers intervene and sell different types of beans to the final consumer in informal marketplaces or markets. Both buyers and sellers (in this case the producer) can be natural persons and their families, agricultural companies and cooperatives (various caponese), wholesale and retail companies, companies from other sectors of the economy, service providers and governments. The final consumer can also support the sale. In this regard, Rocha (2010) mentioned that marketing should facilitate and provide answers to the economic problems of "what" and "how much" to produce, "when", "how" and "where" to distribute the products, and in what "form".

According to this study, very much bean comes from the province of Huila, which is an aspect to be analysed from the market point of view. Regarding this situation, SVagabundo (2014) states that the extension of the market depends on the dispersion of its consumers. However, for two regions (Lubango-Cunene) to be integrated into a single market, there must be the possibility of communication for potential buyers and sellers to maintain contact that allows the transfer of ownership of goods. However, for many products the incorporation of different regions into the same market is limited by transport costs. This is due to the fact that trade between regions will only occur if local prices in different regions differ by an amount greater than the cost of transport. Otherwise, sellers will not be paid for placing their goods in the region of purchase.

The distribution system is composed of all the functions and actions that define the relationship between the producer and its intermediaries. The greater or lesser success depends on the level of integration and cooperation between the parties involved (SILVA, 2010). Therefore, Kotler and Armstrong (1998) consider that the greater or lesser success of the distribution system depends basically on the decisions about channel alternatives and the level of integration and cooperation between the parties involved. Thus, they define a distribution channel as "a set of interdependent organisations involved in the process of making the product or service available to the final consumer".

4.1.2 Establishments selling beans

When analysing the number of bean sellers in both markets, it was found that there is a difference between the types or varieties they sell, being higher in the informal market of Oshomokuyo (see figure 3).

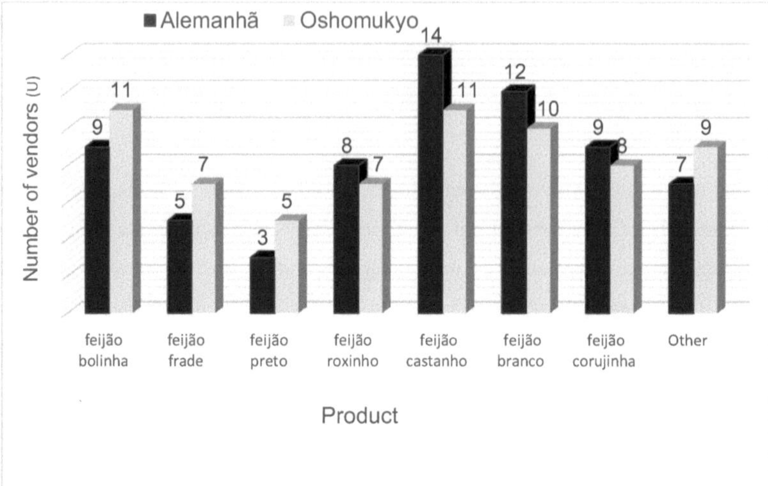

Figure 3. Behaviour of bean presence in informal markets

It was found that there is a varied supply of products, but interviewees commented that the most purchased by customers/consumers in Germany are brown and white beans, with black and black beans being the least purchased. This result may be due to consumption patterns or a cultural aspect of the location. This result is in line with Singh (2017), who states that brown beans are a staple food component of both rural and urban populations in the regions and are therefore highly consumed.

In the Oshomokuyo informal market the results were almost similar, but the supply of beans was found to be higher compared to the other market studied. It was difficult in the research process to count the quantities of beans held by each vendor and all the genotypes sold in the large quantities that exist. Despite the differences in offerings in the two markets, the prices in both markets were found to be the same.

According to the low presence of black beans in the markets, Silva and Oliveira (2013) comment that some types of beans are generally underproduced due to low consumption and accepts dog, being distributed in several points of sale, but their presence in the market is very abundant.

The price of beans for 100% of the buyers shows seasonality in the second half of the year. In the months of January to April the price is better, which may indicate, initially, that the immediate sale of the bean production may be the best option for the producer who cannot always wait for the market signal.

In this regard, Andrade *et al.,* (2010), state that threshed beans, i.e. in bulk, offer attractive prices for consumers and are an important purchasing option.

4.1.3 Bean supply-demand displacement in local markets

According to the interviews conducted (see Table 1) in the German and Oshomokuyo markets, 92% of the consumers surveyed stated that the bean offerings are good, but are not sold at different prices because all sellers agree to unify the price throughout the market, which means that sales levels do not vary much.

Table 1. Analysis of the effects of the simultaneous displacement of demand and supply in the two local markets

Displacement	Increased demand	Constant demand	Demand reduction
Increased supply	Price? Quantity ↑	Price ↓ Quantity ↑	Price↓ Quantity?
Constant supply	Price ↑ Quantity ↑	No change	Price↓ Quantity ↓
Supply reduction	Price ↑ Quantity?	Price ↑ Quantity ↓	Price? Quantity ↓

It is important to note that in the real world there are simultaneous shifts in the supply and demand curves in both markets. Therefore, an increase in supply is not accompanied by a reduction in prices, because there may also have been an increase in demand. Table 1 above presents the possible effects of the change in supply and demand on the market equilibrium price and quantity.

The price of the good or service is the main information available for decision-making, with buyers wanting to pay the lowest possible price and sellers wanting to charge the highest possible price for everything in the Lebanese city of Ondjiva that was not part of this study. . In general, chaos does not occur, there are relatively long periods of time with some price stability, situations where prices oscillate around a stable value, called market equilibrium (ME) by economic theory.

Thus, the model proposes that quantities demanded and supplied will adjust to a price level that meets the objectives of buyers and sellers (Hall and Lieberman, 2013; Pinho *et al.*, 2014).

The main variable is the price of the good or service. The supply model predicts that when the price of a good increases and all other variables remain unchanged, the quantity supplied of that good increases. This occurs because the higher price increases profitability, which makes sellers interested in increasing their supply of beans.

4.2 Identification of the main problems affecting the commercialisation of beans in the informal markets of Cunene.

4.2.1 SWOT Matrix

Taking into account the internal and external diagnosis represented in table 2, the management tool Matrix (FOFA) was used for beans in the two informal markets studied and to expose strategies.

Because the opportunities, weaknesses and threats included in the diagnosis are specific to the business process here in the province, these variables are unique in their functioning and results, and should not be generalised.

Table 2. Strengths, weaknesses, opportunities and threats

Strengths	A traditional and highly appreciated product throughout the province; Hardy and hardy, it is an important source of energy, with a low fat content; It is adaptable to different types of climate and soil and can be grown alone, in intercropping or intercropping; Research, financing and production support agencies encourage innovation in the production chain, overcoming challenges related to new pests, increasing productivity and necessary investments; The existence of three crops makes it easier to change planting intentions throughout the year, which can influence prices.

Weaknesses	Product with short shelf life; Concentration of production in beans, well accepted domestically, but little produced for export; Family production is low-tech and low-professional, using homemade seeds, which degenerates the varieties planted and facilitates contamination by pathogens and mechanical damage, as a result of the low capitalisation of producers and deficient technical assistance.
Opportunities	Research into new products prepared or semi-prepared with beans, reducing preparation time, such as hamburgers, (gluten-free) flour for making bread, biscuits and pasta; Research into new varieties that are earlier, more productive or with larger beans (European preference) to avoid concentrating on beans and to offer more options in the market; Development of traceability systems, due to the change in the profile of the consumer, who is more demanding and seeks to know the origin and conditions in which the product is grown; Marketing campaigns promoting the nutritional qualities of beans.
Threats	Climate change tends to make extreme events, such as droughts or floods, more intense and have shorter cycles of occurrence. Emergence of new pesticide-resistant pests and diseases.

Situations encountered in both places, presupposing similar reactions, e.g. a weakness or threat in elaquid Germany will not necessarily be the same for Oshomokuyo.

In this sense, Steiner George (1995) states that improvement proposals should be aimed at reducing weaknesses, strengthening and maintaining strengths, by looking for suitable opportunities for capabilities, such as providing a defence against external threats. In another paper, Thompson and Strickland (2001) published that accepted improvements should be congruent with the conditions or means of operation of the marketing process, to grow its skills and resources.

A series of strengths and opportunities were found to confront the process, which, if taken advantage of, could reverse the development of the agricultural trade of beans in the province, with benefits for the family economy, but specifically in the improvement of agricultural management, availability of mechanisms and sources of acquisition of inputs to improve infrastructure and, in the long term, improve prices, as well as the availability of beans in the markets. A group of threats were identified that over time imply a setback in this field, which can be reversed by achieving unisectoriality, so that the institutions involved and present in the province can work towards a common goal.

4.3 Proposal of actions for the commercialisation process of beans (*Phaseolus vulgaris* L.) in informal markets in the Commune of Ondjiva in the Municipality of Cuanhama.

4.3.1 Proposal for action

The actions described above should be carried out in different ways, but the most efficient is through governmental action projects with a clear intentionality in their execution and implementation, or also through agricultural extensionism, an aspect that is little explored in the province and also limited (see table 3).

Related to the above, the identification of weaknesses constitutes the first step towards the improvement of the living conditions of the population through a defined, determined and repaired action among local and public actors based on the use of business management capacities and the creation of an innovative territorial environment to implement the proposed actions.

Table 3. Proposed actions for the commercialisation process of French beans (*Phaseolus vulgaris* L.)

WEAKNESS ENCOUNTERED	ACTION	FUTURE PERFORMANCE	MONITORING INDICATORS		
			Weather	**Responsible**	**Control**
Product with short shelf life.	Ensure storage of bean volumes during harvest periods.	Increased storage period; Increased presence of beans in times of low production.	Once a year;	Vendor; Farmers.	Control, Periodic review by MINAGRI experts.
Objective:	Increase the capacity to store beans in grain.				
Concentration of bean production, well accepted domestically, but little produced for export.	Develop a marketing strategy that foresees the distribution of beans according to the different seasons of the year.	To achieve an equitable presence of commercial quantities of beans throughout the year.	annually	IDA, MINAGRI	check, Periodic review by MINAGRI experts.
Objective:	Increase customer satisfaction by increasing the presence of beans in the market throughout the year.				
Inadequate hygienic conditions	Use packaging and improve the	Maintain hygienic and sanitary conditions	Twice a year.	Salesperson, MINAGRI	check,

	presentation of beans in informal markets to preserve their safety.	when selling beans in markets.			Periodic review by MINAGRI experts.
Objective:	Improve the quality and hygienic-sanitary conditions of beans traded on the market.				
Family production is low-tech and low professionalisation, using homemade seeds, which degenerates the varieties planted and facilitates contamination by peptogens and mechanical damage, as a result of poor technical assistance.	Design development projects for the introduction of new technologies. Design and manage a training programme.	Increase the use of technology and the professionalisation of technical assistance.	twice a year	MINAGRI, extension agents, IPO.	Report and reports.
Objective:	To increase technology in the production and professionalisation of beans with a view to improving the culture in the locality.				

High number of competitors	Competing in a market where there are other bean sellers. Offer beans at different prices.	Maintain stability in earnings. Raise sales levels.	Weekly	Vendor; Peasant.	check, Periodic peer review.
Objective:	Secure bean sales levels, although competition is high.				

V. CONCLUSIONS

a) the marketing of frijols if you work under the sales model: producer-copier or intermediary-wholesaler and consumer

b) In the German and Oshomokuyo markets, 92% of consumers surveyed stated that the food offerings of beans are good, but are not sold at different prices, and in Oshomokuyo there is a greater variety of beans on sale compared to Germany.

c) The situations found in FOFA in both places presuppose similar reactions, but it was found that a weakness or threat in Germany here will not necessarily be for the Oshomokyo market.

d) The actions outlined above should be implemented in different ways, but the most efficient way is through government action projects with a clear intentionality in their implementation.

VI. RECOMMENDATIONS

a) To deepen the study of the bean marketing process.

b) Implement the proposed actions set out in the work.

c) Socialise this work in different settings.

✓ VII. BIBLIOGRAPHICAL REFERENCES

- ✓ Agribusiness (2014). Networks of cooperation of the Agribusiness Technologies and Services Branch. Luanda.
- ✓ Almeida, ALG DE, Alcântara, RMCM DE Nobrega, RSA, Nobrega, JCA, Leite, LC, Silva, JAL (2010). Productivity of cowpea cv BR 17 Gurguéia inoculated with symbiotic diazotrophic bacteria in Piauí. Brazilian Journal of Agricultural Sciences, v. 5, no. 3, p. 364 - 369.
- ✓ Andrade, FN, Rocha, M. de M., Gomes, RLF, Freire Filho, FR, Ramos, SRR (2010) Estimates of genetic parameters in cowpea genotypes evaluated for fresh beans. Revista de Ciencias Agrícolas, Fortaleza, v. 41, no. 2, p. 253-258, April/June.
- ✓ Arbage, AP (2016). Fundamentos de la Economía Rural. Chapecó: Argos, 20pp.
- ✓ Barros, G. (2007). Economics of agricultural marketing. Centre for Advanced Studies in Applied Economics - CEPEA. University of São Paulo - USP.
- ✓ Barros, GSAC, JGMartines Filho, (2017). "Price Transmission and Marketing Margins of Agricultural Products" in DELGADO, GC, JG GASQUES and CM VILLA VERDE (org.), Agriculture and Public Policies. Serie IPEA núm. 127, Brasilia-DF.
- ✓ CGA (2010) Consulate General of Angola. Kwanza Norte, Capital: N'dalatando. Retrieved May 15, 2021, from the website of: Consulate General of Angola: http://www.consuladodeangola.org/index.php?Itemid=168&id=188&option=com_ content&tas k=view
- ✓ Castiñeiras L. (2011). In situ management and conservation of farm-grown plant genetic resources in Cuba. Agricultura Orgánica. Havana. Cuba
- ✓ DIARIO DIARIO DE LA REPÚBLICA (2014). Estrategia Nacional de Comercio y Emprendimiento Rural. Gazette of the Republic, 2
- ✓ Diniz, AC (2018). Angola's physical environment and agricultural potential. Instituto de Cooperacion Portuguesa. Lisbon.
- ✓ Eterno, PV and Elenor, AAW (2015). Analysis of bean marketing channels in production centres in the Eastern Region of the State of Goiás Revista Cojuntura Econômico Goania, June, No:33,16-25pp.
- ✓ FAO. Food and Agriculture Organization of the United Nations. FAOSTAT. (2019). Available at: <http://faostat.fao.org/default.aspx>. Accessed on: 26 December. 2021.

- FAO. Food and Agriculture Organisation of the United Nations. FAOSTAT (2020). Available at: <http://faostat.fao.org/default.aspx>. Accessed on: 21 January. 2022.
- Ferreira, CM, Peloso, MJ Del, Faria, LC (2012). Beans in the national economy. Santo Antonio de Goiás: Embrapa Rice and Beans. 47p. (Embrapa Rice and Beans. Documents, 135).
- FIEP (Federation of Industries of the State of Paraná) (2017). Programme to increase sales of products from Paraná. FRIJÓLES - Version 1.0. May. 2006. Available at: <http://www.fiepr.org. br/fiepr//conselhos/agroindustria_alimentos/uploadAddress/Relat%C3%B3rioFeij %C3%A3%Ao0506.pdf>. Accessed on: 01 January. 2021.
- Figueiredo, A., Prescott, E., Melo, MF de. (2004). Integration between the household market and the retail market. Brasilia: Universal. 196p.
- Fisher, L. and Navarro, A. (2018). Introducción a la Investigación de Mercados, Ed. McGraw Hill, 2nd Ed, Mexico, 162 pp.
- Frank, HF (2011). Microeconomics and behaviour. São Paulo: McGraw-Hil. 3rd edition. pp. 300-379.
- Garófalo, G. de L; Carvalho, LCP de. (2014). Teoría microeconómica. São Paulo: Atlas SA 2nd ed. pp. 170-260.
- Gepts, P., Debouck, DG (2015). Origin, domestication and evolution of the common bean (*Phaseolus vulgaris* L.). In: SCHOONHOVEN, A. van; VOYSEST, O. Common bean research for crop improvement. Wallingford: CAB International; Cali: CIAT. pp. 7-53.
- GonzálezM, (2018). Fungal diseases of beans. Editorial Científica Técnica. La Habana. Cuba
- Guzmán, ES (2015). Agroecology and sustainable rural development. In: AQUINO, AM; ASSIS, RL Agroecology: principles and techniques for sustainable organic agriculture. Brasilia, DF: Embrapa. pp. 101-131.
- Hall, M., Lieberman, M. (2013). Introduction to futures and options markets. São Paulo, Bolsa de Mercancias e Futuros (BM&F) - Cultura Editores Associados, 2nd expanded edition, 19pp.
- Hernández, SR, Colinas, C. and Baptista, P. (2003). Research Methodology, 684pp.
- José, E. (2015). The gurus' guide: the concepts of best business practices. São Paulo: Campus. pp. 189-204.

- ✓ Angola Review (2021). Bean production with a deficit of 350,000 tonnes. Informative article. Available: https://www.jornaldeangola.ao/ao/noticias/producao-de-feijao-com-defice-de-350-mil-toneladas/consultado: 11/07/2022
- ✓ Provincial Liquidation Board (2015). Application of the method to the planning of an agricultural enterprise located in the village of Luinga, Ambaca municipality, Cuanza Norte district, Angola. Reordering, pp. 3-7.
- ✓ Kaplan, L. (2016). Archaeological Phaseolus from Tehuacán. In: BEYERS, D. (Ed.). The prehistory of the Tehuacán Valley: environment and subsistence. Austin: University of Texas. v. 1, p. 201-212.
- ✓ Kaplan, RS, Cooper, R. (2018). Cost and performance: manage your costs to be more competitive. São Paulo: futuro. 366 pp.
- ✓ Kotler, Philip (2000). Principles of marketing. São Paulo: Prentice Hall.
- ✓ Ledesma, A. M. (2017). Agribusiness, enterprise and entrepreneurship. 2nd ed. Buenos Aires: El Ateneo. 266 pp.
- ✓ Momot. (2013). Strategy for the Valorisation of Local Agricultural Production. Final Report of the GEVPAL Working Group. Ministry of Agriculture, Sea, Environment and Spatial Planning. Angola
- ✓ Mendes, Judas Tadeu Grassi; PADILHA JUNIOR, João Batista (2007). Agribusiness: an economic approach. São Paulo: Pearson Prentice Hall,
- ✓ MINADER-FAO. (2003). Review of the agricultural sector and food security strategy to define investment priorities (TCP/ANG/2907) ± Agricultural Production Systems. Working Paper No. 07. Preliminary version for comments. Accessed 13 May 2021, on the website of: Ministry of Agriculture and Rural Development of Angola:http://www.minader.org/pdfs/fomento/volume_iii/sistema_producao_agricola.pdf
- ✓ MINADER, (2020) Construtora Norberto Odebrecht SA and Sondotécnica Engenharia de Solos SA Rehabilitation Project of the General Matala Conductor Canal ± Capelongo. Province of Huila ± Municipality of Matala.
- ✓ Moreira, I. (2021). Angola Agriculture, Natural Resources and Rural Development. Volume I. ISA-Prensa. Lisboa.
- ✓ Muñoz, G. and Singh, S. (2003). Comparative studies of sources of common Bacteriosis resistance available in different Phaseolus species and genetic progress through interspecific crosses and gene pyramiding. In SP Singh & O. Voysest,

(Eds.), Workshop on bean breeding for the 21st Century: Basis for a strategy for Latin America. CIAT. Cali. Cali. Colombia.

- Navolar, T. S., Rigon, A. S., Philippi, S. J. M. (2010). Dialogue between agroecology and health promotion. Brazilian Journal of Health Promotion, Fortaleza, v. 23, no. 1, p. 69-79.
- Neves, MF (Org.) (2010) Economics and management of agri-food enterprises: food industry, input industry, agricultural production, distribution. São Paulo: Pioneer. ch. 3, pp. 39-60.
- Neves, MF, Castro, LT (2017). Marketing and strategy in agribusiness and food. São Paulo: Atlas, 365 p.
- NEVES, Marcos Favá; Chaddad, Fabio Ribas; Lazzarini, Sergio Giovanetti (2013). Management of food companies. São Paulo: Thomson Pioneer Learning, 14pp.
- Pinho, Diva Benevides; VASCONCELLOS, Marco Antônio Sandoval de (Org.) (2014). Manual de economia. 5. ed. São Paulo: Saraiva, 33pp.
- Piza, CT, RW Welsh, (2018). Introduction to marketing analysis. Manuals Series No. 10. Department of Economics - ESALQ/USP, Piracicaba-SP.
- Porter, Michael E. (2006) Competitive strategy: techniques for analysing industries and competition. Rio de Janeiro: Campus, 32pp.
- Ramírez, R. J. L. (2007). Course material: Gestión Estratégica, Maestría en Ciencias Administrativas, IIESCA UV, Mexico.
- Rivas, G. (2007). Economics of agricultural marketing. Centre for Advanced Studies in Applied Economics - CEPEA. Universidad Agustino Neto, 23-35pp.
- Rocha, M. de M. (2010) Cowpea for consumption as fresh beans. Agrosoft Brasil, 11pp. Available at: http://www.agrosoft.org.br/agropag/212374.htm. Accessed on: 5 July. 2021.
- Rocha, MM, Freire-Filho, FR, Ribeiro, VQ, Carvalho, HWL, Belarmino-Filho, J., Raposo, JAA, Alcântara, JP, Ramos, SRR, Machado, CF (2007). Adaptability and production stability of semi-erect cowpea genotypes in the Northeast Region of Brazil. Brazilian Agricultural Research, 42: 1283-1289.
- Rodríguez, V.J. (2000). Administración con Enfoque Estratégico. Editar. Trillas, Mexico.
- Sabourin, E. (2010). Multifunctionality and non-market relations: common resource management in the Northeast. CRH Notebook, n. 22, vol. 56, p. 151-169.

- ✓ Scubla L. (1985). Logics of reciprocity. Paris, Ecole Polytechnique, Cahiers du CREA n. 6, 283p.
- ✓ SENAES, National Secretariat of Solidarity Economy (2013). Solidarity Economy in Development Programme, Brasilia: SENAES-MTD.
- ✓ Senaes. (2017). Draft Law on Fair and Solidarity Trade (CJS). Brasilia: SENAES-MTD.
- ✓ Silva, EF (2018). Bean cultivation at the rural scale. São Paulo: Escala, n. 5, p. 10 - 17.
- ✓ Silva, I. (2010). The marketing process in the company Fazenda Sapucaia, municipality of Santa Isabel do Pará. Administration and Technology, v. 1, no. 1.
- ✓ Silva, PSL, Oliveira, CN (2013). Green and mature bean yields of cowpea cultivars. Brazilian Horticulture, Brasilia, v. 11, no. 2, p. 133-135.
- ✓ Cantante, P. (2016). Introduction to the Solidarity Economy. São Paulo: Perseu Abramo, 127p.
- ✓ Singh, BB (2017). Recent progress in cowpea genetics and breeding. Acta Horticultura, The Hague, n. 752, pp. 69-76, 2007. Proceedings edition of the International Conference on Indigenous Vegetables and Legumes, Hyderabad, India, September. Available at: http://www.actahort.org/books/752/752_7.htm
- ✓ Socorro, M and Martín D. (2009): Granos, Editorial Pueblo y Educación. Havana. Cuba
- ✓ Steiner-George, A. (1995). Strategic Planning. Editar. CECSA, Mexico.
- ✓ Tamayo, M. (2003). El proceso de investigación científica, 440 pp., Limusa-Noriega Editores. Mexico, 43pp.
- ✓ Thompson, L. and Strickland, A. (2001). Strategic Management. Editar. MacGraw-Hill, Colombia.
- ✓ Unifeijão (2011). Bean classification. Available at: <http://www. unifeijao.com.br/telas/class_feija.php>. Accessed on: 29 Dec. 2021.
- ✓ Wander, AE (2007). Production and consumption of beans in Brazil, 1975-2005. Información Económica, São Paulo, v. 37, no. 2 February.
- ✓ Wander, AE (2014). Irrigated bean cultivation in the northwestern region of Minas Gerais. Santo Antônio de Goiás: Embrapa Rice and Beans. Available at: <http://sistemasdeproducao.cnptia.embrapa.br/FontesHTML/Feijao/FeijaoIrrigado NoroesteMG>. Accessed on: January. 2021

VIII. ATTACHMENTS

Annex 1.

Bean genotypes on the German market.

Oshomukyo Consultation Day

Survey Day in Germany

ANNEX 2

Name of respondent:

Age: _____. Lugar: ________________________________. Date: __________.

Main channels used by photocopiers to sell beans to

() Partnerships

() Praxas

() Intermediaries

() Supermarkets, shops

() Direct sale at the production site.

() Otro: ____.

Main varieties you sell?

In what unit of measurement is the product sold?

In kilograms ______ in pounds ______ tonnes ______ other ______

What would help you sell more?

Who buys more products to consume?

Women _______________ men _______

When do prices increase?wYou?

How is it traded? beans to?

How many kg do you sell per day?

which is the beanThey are bought less and less?

MIX
Papier aus verantwortungsvollen Quellen
Paper from responsible sources
FSC® C105338

Printed by Books on Demand GmbH, Norderstedt / Germany